MW00846650

SPACE SCIENCE

MOONS

BY BETSY RATHBURN

BELLWETHER MEDIA • MINNEAPOLIS, MN

Are you ready to take it to the extreme? Torque books thrust you into the action-packed world of sports, vehicles, mystery, and adventure. These books may include dirt, smoke, fire, and chilling tales. **WARNING**: read at your own risk.

This edition first published in 2019 by Bellwether Media, Inc.

Library of Congress Cataloging-in-Publication Data

Names: Rathburn, Betsy, author.
Title: Moons / by Betsy Rathburn.
Description: Minneapolis, MN : Bellwether Media, Inc., [2019] | Series:
   Torque: Space Science | Audience: Ages 7-12. | Includes bibliographical
   references and index.
Identifiers: LCCN 2018001136 (print) | LCCN 2018008541 (ebook) | ISBN
   9781681036014 (ebook) | ISBN 9781626178601 (hardcover : alk. paper)
Subjects: LCSH: Satellites–Juvenile literature.
Classification: LCC QB401.5 (ebook) | LCC QB401.5 .R38 2019 (print) |
   DDC 523.9/8–dc23
LC record available at https://lccn.loc.gov/2018001136

Editor: Rebecca Sabelko   Designer: Andrea Schneider

Printed in the United States of America, North Mankato, MN.

TABLE OF CONTENTS

ONE SMALL STEP

A **lunar module** touches down on a dark, rocky surface. Two **astronauts** peer through a window. Earth is a small circle in the sky. The astronauts are 238,855 miles (384,400 kilometers) from home!

With one step, Neil Armstrong becomes the first person to set foot on the Moon. The rocks and dust collected will be studied for years to come!

NEIL
ARMSTRONG

WHAT ARE MOONS?

Moons are large bodies of rock in outer space. They are also called **satellites**. Many are covered in deep **craters**. These were created by **impacts** with other objects.

Moons may have tall mountains. Some were formed by impacts. Others come from **volcanoes**. The tallest moon mountains in the solar system are more than 10 miles (16 kilometers) high!

FUN FACT

A VERY BIG VOLCANO

The largest known volcano in the solar system is on Mars. It is called Olympus Mons. This volcano is more than 15 miles (24 kilometers) tall!

Olympus
Mons

CRATER

PHOBOS

No two moons are alike. Many are shaped like **spheres**. Others may be bumpy or potato shaped. Some even have their own **atmosphere**!

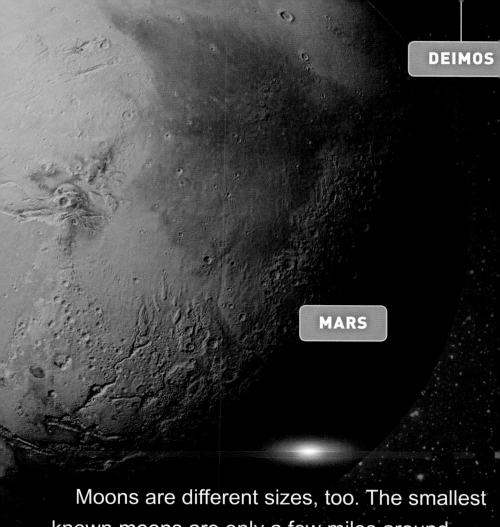

DEIMOS

MARS

Moons are different sizes, too. The smallest known moons are only a few miles around. Other moons are bigger than planets!

HOW DO MOONS FORM?

Millions of years ago, the solar system swirled with dust and gas. These materials came together to form some of the solar system's moons.

Other moons came from clumps of debris passing through the solar system. The gravity of planets pulled the debris into orbit. This debris became moons!

EARTH'S MOON

Distance from Earth:
   238,855 miles (384,400 kilometers)
Length of orbit: 27 days
Size: 6,784 miles
   (10,917 kilometers) around
Highest point:
   35,387 feet (10,786 meters)
Temperature range:
   -280 to 260 degrees Fahrenheit
   (-173 to 127 degrees Celsius)

ome moons were created from impacts
je objects struck planets and sent
ıks of rock into space. Gravity kept
debris in orbit around the planets.

Astronomers believe Earth's moon was formed from an impact. An object the size of Mars hit Earth. The materials thrown into space joined and became Earth's moon!

WHERE ARE MOONS FOUND?

There are hundreds of known moons in the solar system. Most of these moons orbit planets. Some planets, like Mercury and Venus, have no moons. Others, like Jupiter and Saturn, have more than 50!

Sometimes, moons can travel around other large objects. More than 200 moons have been found in orbit around asteroids!

FUN FACT

MEGA MOON

The largest moon in the solar system orbits Jupiter. It is called Ganymede. Ganymede is bigger than the planet Mercury!

JUPITER

JUPITER'S MOON,
GANYMEDE

ILLUSTRATION
OF EXOMOON

There are likely many moons beyond
the solar system, too. In 2017, astronomers
spotted what may be the first exomoon
seen from Earth.

PLANETARY MOONS

Mercury: 0   **Jupiter:** 53

Venus: 0   **Saturn:** 53

Earth: 1   **Uranus:** 27

Mars: 2   **Neptune:** 13

With more research and advanced **telescopes**, astronomers will likely discover more exomoons in the future. Moons are found all over the universe!

WHY DO WE STUDY MOONS?

Astronomers study moons to learn more about the universe. They look at the size and shape of moons. This tells scientists where the moons came from. It also helps them discover what moons are made of.

Studying moons helps scientists learn how the solar system formed. It gives clues to how life on Earth started!

FUN FACT

FLYBY MOON

In 2015, a small asteroid called BL86 flew by Earth. It had its own moon!

The Hubble Space Telescope helps scientists study moons. In the future, even more powerful telescopes will help do this job. **NASA** plans to launch the James Webb Space Telescope in 2020. With more power to see far-off worlds, astronomers on Earth may discover even more moons!

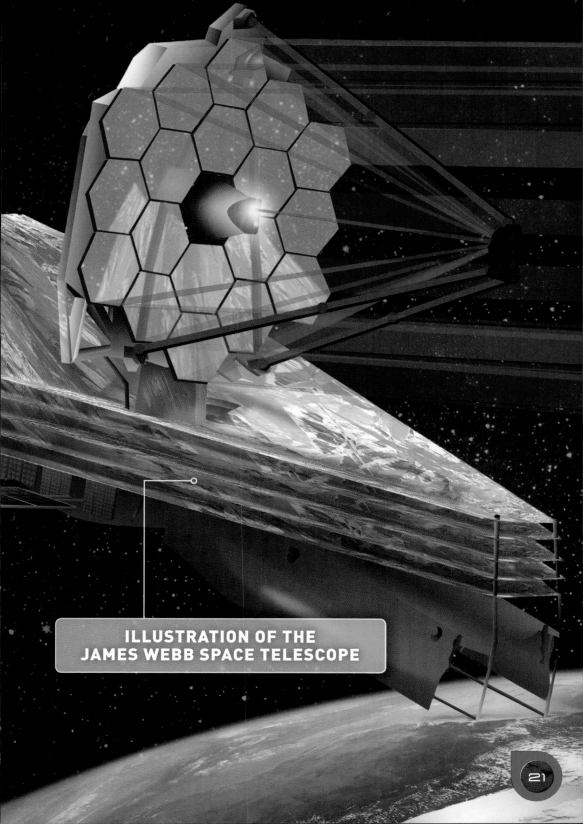

ILLUSTRATION OF THE
JAMES WEBB SPACE TELESCOPE

GLOSSARY

astronauts–scientists who explore and study space

astronomers–people who study space

atmosphere–the gases that surround planets and some moons

craters–deep holes in the surface of a moon or other object

debris–leftover materials

exomoon–a moon outside of Earth's solar system

gravity–the force that pulls objects toward one another

impacts–events in which objects hit one another

lunar module–the part of a spacecraft that lands on a moon's surface

NASA–National Aeronautics and Space Administration; NASA is a U.S. government agency responsible for space travel and exploration.

orbit–a complete movement around something in a fixed pattern

satellites–objects that orbit planets and asteroids

spheres–ball-shaped objects

telescopes–instruments used to view distant objects in outer space

volcanoes–vents that let out hot rocks and steam

TO LEARN MORE

AT THE LIBRARY

Hansen, Grace. *The Moon*. Minneapolis, Minn.:
Abdo Kids, 2018.

Hubbard, Ben. *Neil Armstrong and Getting to the Moon*.
Chicago, Ill.: Heinemann Raintree, 2016.

Troupe, Thomas Kingsley. *Apollo's First Moon Landing*.
North Mankato, Minn.: Picture Window Books, 2018.

ON THE WEB

Learning more about moons
is as easy as 1, 2, 3.

1. Go to www.factsurfer.com

2. Enter "moons" into the search box.

3. Click the "Surf" button and you will see a list of
   related web sites.

With factsurfer.com, finding more information is just a

INDEX

The images in this book are reproduced through the courtesy of: Milissa4like, front cover, pp. 3, 4,
8, 10, 12, 14, 16, 18, 20, 23 (graphic); Dotted Yeti, front cover (moon), pp. 8-9, 16 (exomoon), 1
(exoplanet); Alan Uster, front cover, pp. 2-3 (Earth/Moon); kvsan, p. 2; INTERFOTO/ Alamy,
pp. 4-5; HelenField, pp. 6-7; ixpert, p. 7 (Earth); NASA/JPL/USGS, p. 7 (inset); sdecoret, pp. 10-
REDPIXEL.PL, pp. 12-13 (Earth), 13 (Moon); Triff, pp. 12-13, 14-15; Vadim Sadovski, p. 15 (Jupiter,
Ganymede); frantic00, pp. 18-19; edobric, pp. 20-21 (telescope); xtock, pp. 20-21 (Earth).